The divisors and the multiples

Basic concepts of mathematics

Zid mohamed

ISBN-13: 978-1539962090

ISBN-10: 1539962091

Zid MOHAMED

The divisors and the multiples

Basic concepts of mathematics

Huda Editions

Dedication

I dedicate my modest manuscript to all scholars and researchers who did not skimp on resources. They have given substance by sacrificing himself at the expense of their health, to present us always new about theories, theorems, and applications in this vast and exciting field that is mathematics.

All due respect, our masters and teachers in all school levels, as well as my strong encouragement to all students, and enthusiasts passionate exact sciences and mathematics in particular.

1 Préface

If the construction of a gigantic edifice rests on a very solid foundation, all the sciences without exception, based on the large-scale mathematical sciences.

Student in math class high schools, and Professor of technology, I was passionate about mathematics since my young age and I will remain.

Formerly, my refuge and my distraction was always looking any cost solutions mathematical exercises, and you can not imagine the joy that overwhelms me in these moments, and it was always a challenge for me.

No doubt the will and self-sacrifice in seeking solutions math problems inevitably lead to strengthening the human mind and the same, it will develop a great personality and a mind at all events.

My own experience in this area, I taught my students, thank you god, they are heavily exploited, and for their studies and their future.

My strongest advice is the permanent removal of the minds of children and parents misconceptions, like what, the mathematics are complex and exacting, law, on the contrary, it is as and extent that the child applies to individually solve its exercises, it will take taste and never will relax, that I promise you and I remain totally convinced.

My dearest wish remains that everyone, will become aware of the paramount importance of the mathematical sciences, and work accordingly to the base, in order to achieve its objectives.

2 Prologue

In my series entitled "Mathematics in basic notions," I present this little school booklet, where all my attention was focused around a key topic of mathematics, and I've always considered its base and its main pillar this is none other than the divisibity.

Given the extreme importance of this subject, I started with the simple division and divisibility of natural numbers, to facilitate the understanding and assimilation is from this simple operation, and only this one that they will assimilate all the subject, and we will have no difficulty to understand and solve all the exercises, to the upper level, which is the university.

Table of contents

3 Euclidean division

3.1 Application

- If a person distributes sweets among 10 two (02) children, each obtains effective May 5 candy, and is left with nothing.

We can say that this person has divided its 10 candy by 2, which gives (05) candy for everyone, and nothing left him, ie a zero 0.

Operation is performed as follows:

$$
\begin{array}{c|c} 10 & 5 \\ \hline 0 & 2 \end{array} \qquad \begin{array}{c|c} D & d \\ \hline r & q \end{array}
$$

- The number 10 is called the dividend D

- The number 2 is called the divisor d

- The number 5 is called the quotient q

- The number 0 is called the remainder r

This simple operation is called division.

3.1.1 Definition
The Euclidean division is an arithmetic operation of dividing any number, the dividend by another, the divisor, and the result is a natural number, the quotient and a natural number, the rest.

We writes : $D \div d = q$ rest r or $D / d = q$ rest r we will have:

$$
\boxed{D = (q \times d) + r}
$$

3.1.2 Applications exercices

- $1 \div 1 = 1$ rest $0 \leftrightarrow 1 = (1 \times 1) + 0$

- $2 \div 2 = 1$ rest $0 \leftrightarrow 2 = (2 \times 1) + 0$

- $3 \div 2 = 1$ rest $1 \leftrightarrow 3 = (1 \times 2) + 1$

- $4 \div 2 = 2$ rest $2 \leftrightarrow 4 = (2 \times 2) + 0$

- $5 \div 3 = 1$ rest $2 \leftrightarrow 5 = (1 \times 3) + 2$

- $7 \div 4 = 1$ rest $3 \leftrightarrow 7 = (1 \times 4) + 3$

- $8 \div 3 = 2$ rest $2 \leftrightarrow 8 = (2 \times 3) + 2$

- $9 \div 5 = 1$ rest $4 \leftrightarrow 9 = (1 \times 5) + 4$

- $10 \div 1 = 1$ rest $0 \leftrightarrow 10 = (10 \times 1) + 0$

- $10 \div 2 = 5$ rest $0 \leftrightarrow 2 = (5 \times 2) + 0$

If we denote by **a** the dividend, **b** the divisor, and the quotient **k**, **r** the rest, we will :

$$a = (k \times b) + r$$

11

Note that in the above examples

- Three **(03)** operations where the rest is each time a zero, **r = 0**

- Three operations when the rest is in each case a different number of zero **r ≠ 0**

We deduce from these operations:

-       **1 ÷ 1 = 1** rest **r = 0,** which means 1 divides **1, ou 1** is a divisor of **10**

-       **2 ÷ 2 = 1** rest **r = 0,** in that case 2 divides **2**, ou **2** is a **divisor** of **2**

-       **3 ÷ 2 = 1** rest **r = 1,** it means that **2** does not divide **3**, ou **2** is not a **divisor** of **3**

-       **4 ÷ 2 = 2** rest **r = 0,** it means that **2** divise **4**, ou 2 is a **divisor** of **4**

-       **5 ÷ 2= 2** rest **r = 1,** it means that, **2** does not divide **5**, ou **2** is not a **divider** of de **5**

-       **6 ÷ 2 = 3** rest **r = 0,** which means **2** divides **6**, ou **2** is not a **divider** of **6**

-       **7 ÷ 5 = 1** rest **r = 2,** which means **5** does not divide **7**, ou **5** is not a **divider** of **7**

-       **8 ÷ 5 = 1** rest **r = 3,** it means that **5** does not divide **8**, ou **5** is not a **divider** of **8**

-       **9 ÷ 5 = 1** rest **r = 4,** it means that **5** does not divide **9**, ou **5** is not a **divisor** of **9**

-       **10 ÷ 4 = 2** rest **r = 2,** it means that **4** does not divide **10**, ou **4** is not a **divider** of **10.**

All of these applications are deduced theorems below:

3.1.3 Theorems

1 - Theorem 1

A natural number is different from zero $a \neq 0$, is divisible by an integer **b**, if and only if, the rest of this division is zero, $r = 0$, $r \in N^*$

2 - Theorem 2

They say the natural number b is a divisor of the natural number a, if and only if, there is an

integer $k \neq 0$ satisfying the equation $a = (k \times b) + r, r = 0$

Reciprocally :

3 - Theorem 3

If there is an integer **b** is a natural number **a**, and the natural number $k \neq 0$, and the natural

number r and satisfying the equation $a = (k \times b) + r, r = 0$, with $k \neq 0$, we say that **b** is **a**

divisor of **a**

4 The odd and even numbers

4.1 Even numbers

4.1.1 Definition

The pair is having pairs of pairs in the various cases.

□ Examples of applications

Take for example the ten (10) digits are:

1, 2, 3, 4, 5, 6, 7, 8, 9, 10, one notices that there are even numbers and odd numbers, in this case, we look for the mathematical rules that differentiate them from each other, there are two theories:

If you take one, it is unity, there is only 1

The figure **2** is composed of two units 1 and 1 or 1 + 1, that is to say two peer 1, ie, it is divisible by 2, because the remainder r = 0, $2 \div 1 = 2$, r = 0

$2 \div 2 = 1$ rest 0, so the figure 2, is an even number.

Number 3, if we divide the peer 2, there will be 2 + 1 or 1 + 2, there is only one pair, and the rest is 1, r = 1. It is indivisible by 2 so this is an odd number.

The figure **4** is composed of two 2-digit and 2 or 2 + 2, that is to say, two pairs of two, and secondly, it is divisible by 2, because the remainder r = 0, $4 \div 2 = 2$, still 0, so the number 4, is an even number.

5, we can write 4 + 1 or 2 + 2 + 1, that is to say, two pairs of 2 plus 1, that is, it is indivisible by 2 because the remainder is 1, $5 \div 2 = 2$, r = 1 so the number 5 is odd.

Number **6**, we can write 4 + 2 or 2 + 2 + 2, that is to say, three pairs of 2, that is, it is divisible by 2, because the rest r = 0, $6 \div 2 = 3$, r = 0, so the figure 6 is an even number.

The number **7**, we can write 6 + 1 or 2 + 2 + 2 + 1, that is to say, three pairs of 2 plus 1, that is, it is indivisible by 2 because the rest is 1 , r = 1 so the number 7 is odd.

The number **8**, we can write 4 + 4, or (2 + 2) + (2 + 2), that is to say four pairs of 2, that is, it is divisible by 2, because the rest r = 0, $8 \div 2 = 4$, r = 0, therefore, the figure 8 is an even number.

Number **9**, we can write 8 + 1 or (2 + 2 + 2 + 2) +1, that is to say four pairs of 2 plus 1, that is, it is indivisible by 2 because the remainder is 1, r = 1, therefore the number 9 is odd.

The number **10**, we can write 8 + 2, or (2 + 2 + 2 + 2) + 2, that is to say, five pairs of 2, that is, it is divisible by 2, because the rest r = 0, $10 \div 2 = 5$, r = 0, therefore, the figure 10 is an even number.

Note :

1- Note for the number 2, 4, 6, 8, are divisible by 2 and the rest is zero, r = 0, are even numbers.

2- In the case of the numbers 1, 5, 7, 9, they are indivisible by 2 each time the rest is 1, r = 1, have odd numbers.

Generalize:

In the first case, it performs division of natural numbers 2, 4, 6, 8 and 2, and is designated by the k-zero integer, $k \neq 0$ the quotient of this division, we note that the rest is in each case equal to zero, $r = 0$.

Denoting by (a), one of the natural numbers, we have:

$2 \div 2 = 1$ remains 0; $4 \div 2 = 2$, $r = 0$; $6 \div 2 = 3$, $r = 0$; $8 \div 2 = 4$, $r = 0$

If we denote by $k \neq 0$, $k \in N^*$ we will have:

$a \div 2 = k$, $r = 0 \leftrightarrow a = 2k + 0 = 2k$.

In the second case, if we denote by k1, k2, k3, k4, the quotients of division of 1, 3, 5, 7, then:

$1 \div 2 = k_1$, $r = 1$; $3 \div 2 = k_2$, $r2 = 1$; $5 \div 2 = k_3$, $r3 = 1$; $7 \div 2 = k_4$, $r4 = 1$

We deduce the following theorems:

Theorem 1

We say that a natural number is even, if and only if, there is an integer $k \neq 0$, $k \in N^*$, which verifies $a = 2k$, $r = 0$

Theorem 2

We say that a natural number is odd, if and only if, there is a natural number that checks $a = 2k + 1$, k and $r \in N$, $r = 1$

4.1.2 Exercise of application

1- Find all the even numbers and all odd numbers pi n, which satisfy $10 \le n \le 20$, and deduce the set **D** of their non-common divisors

It is the union of two sets, and wrote:

(P) = {10, 12, 14, 16, 18, 20}.

(Pi) = {11, 13, 15, 17, 19}

(P) ∪ (pi) = {10, 11, 12, 12, 13, 14, 15, 16, 17, 18, 19, 20}

Divide each number by 2, we have:

$10 \div 2 = 5$ rest 0 ↔ r = 10 = (5 × 2) 0, k = 5 which checks: 10 = 2k, so 10 is even number.

$11 \div 2 = 5$ remains ↔10 1 = (5 × 2) + 1, k = 5 which checks: 11 = 2k + 1, so 11est odd number.

$12 \div 2 = 6$ remainder r = 0 ↔ 10 = (6 × 2) + 0, k = 6 that checks: 12 = 2k, so 12 is even number.

$13 \div 2 = 6$ remains 1↔ r = 13 = (6 × 2) + 1, with k = 6 which verifies: 13 = 2k + 1, so 13 is an odd number.

$14 \div 2 = 7$ remainder r = 0 ↔ 14 = (7 × 2) + 0, with k = 7 which satisfies: 14 = 2k, thus 14 is an even number.

$15 \div 2 = 7$ remains 1↔ r = 15 = (7 × 2) + 1, with k = 7 which satisfies: 15 = 2k + 1, thus 15 is an odd number.

$16 \div 2 = 8 = 0$ ↔ remainder r 16 = (8 × 2) + 0, with k = 8, which satisfies: 16 = 2k, 16 thus is an even number.

$17 \div 2 = 8$ remains $1 \leftrightarrow r = 17 = (8 \times 2) + 1$, with $k = 8$, which satisfies: $17 = 2k + 1$, so 17 is an odd number.

$18 \div 2 = 9$ remains $r = 0 \leftrightarrow 18 = (9 \times 2) + 0$, $k = 9$ which checks: $18 = 2k$, so 18 is even number.

$19 \div 2 = 9$ remains $1 \leftrightarrow r = 19 = (9 \times 2) + 1$, $k = 9$ which checks: $19 = 2k$, so 19 is odd number.

$20 \div 2 = 10$ remainder $r = 0 \leftrightarrow 10 = (10 \times 2) + 0$, $k = 10$ which checks: $20 = 2k$, so 20 is even number.

Note that the sets (**p**) and (**pi**) are:

p = {10, 12, 14, 16, 18, 20}

$\mathbf{p_i}$ = {11, 13, 15, 17, 19}.

We will also have :

10 = 2 × 5 dividers 10 are: {1, 2, 5, 10}

12 = 2 × (6) = 2 × (2 × 3) × 2 = 6, the dividers 12 are {1, 2, 3, 4, 6, 12}

13 × 13 = 1, the divisors of 13 are {1, 13}

15 = 1 × 15, the dividers 15 are {1, 3, 5, 15}

16 = 1 × 2 × (8) = 2 × (2 × 4) =, the dividers 16 are: {1, 2, 4, 8, 16}

17 = 1 × 17 17 dividers are {1, 17}

18 = 2 × (9) = 2 × (3 × 3) = 6 × 3, so the dividers 18 are {1, 2, 3, 6, 9, 18}

19 = 1 × 19 19 dividers are {1, 19}4 = 2 × 7, the dividers 14 are {1, 2, 7, 14}.

20 = 1 × 2 × (10) = 2 × (2 × 5) = (2 × 2) = 5 × 4 × 5, the dividers 20 are {1, 2, 4, 5, 10, 20}.

We deduce the set **D** of their non-common divisors **D** = {1, 2, 3, 4, 5, 6, 7, 8, 9 10, 11, 12, 13, 14, 15, 16, 17, 18, 19, 20}.

5 the divisibility

The divisibility includes all natural numbers different from zero, $N^* = \{1, 2, 3, 4, 5, 6, 7, 8, 9, 10 \dots .n\}$.

5.1 Definition

If one divides a natural numbe **a**, by an integer **b,** and we get a remainder **r = 0**, we say in this case that **b** divides **a**.

If one divides a natural number, by an integer b, and you get the remainder $r = \alpha$, with $\alpha \neq 0$, we say in this case that the integer b, does not divide a, or the whole a is not a divisor of b.

Example 1

Dividing 8 by 2, we obtain 4, and the remainder is 0, $8 \div 2 = 4$, r = 0, in this case, we say that two divided 8, or 2 is a divisor among all divisors of 8, which are d = {1, 2, 4, 8}.

Example 2

1- seek dividers natural numbers of all, $D_1 = \{2, 3, 4, 6\}$.

Note that 2 divides the natural numbers 2, 4, 6 and does not divide the natural number 3 because,

$2 \div 2 = 1$ remains 0

$3 \div 2 = 1$ remains 1; $4 \div 2 = 2$ remains 0; $6 \div 2 = 3$ remainder 0.

Note that 3 divides the natural numbers 3, 6 and does not divide the natural number 4

$3 \div 3 = 1$ remains 0; $4 \div 3 = 1$ remains 1; $6 \div 2 = 3$ remainder 0

It is noted that 4 divides 4, and does not divide 3; $3 \div 3 = 1$ remains 0; $4 \div 3 = 1$ remains 1;

$6 \div 2 = 3$ remainder 0

Note that 6 divides 6; $6 \div 6 = 1$ remains 0

We deduce all d_1 divisors of all D:

$d_1 = \{3, 4, 6\}$

2 - seek non common divisors of the set D = {2, 3, 4,5, 6,7, 8, 9, 10}.

It is noted that 2 divided 2, 4, 6, 8, and 10, and does not divide 3, 5, 7, 9

3 divides 3, 6, 9 and does not divide 4, 5, 7, 8, 9, 10

4 divides 4, 8 and does not divide 3, 5, 6, 7, 9, 10

5 divides 1, 5, and 10, and divides pas2, 3, 4, 6, 7, 9

6 divides 6, and does not divide 3,4, 5, 7, 8, 9, 10

7 divise1, 7 and does not divide 2, 3, 4, 5, 6, 5, 8, 9, 10

Divise1 8, 8 and does not divide 3, 5, 6, 7, 9

10 divides 1 and 10, and does not divide 3, 4, 6, 7, 8, 9

Denoting by d_2, all non-common divisors of the set D, we have:

d2 = {1, 2, 3, 4, 5, 6, 7, 8, 9, 10}.

Take the number **7** of this set, the dividers of **7,** ther are only **1** and **7** and checking **1 < 7** For they satisfy the equation $\mathbf{a = (k \times b) + r}$, with $\mathbf{r = 0}$ or $\mathbf{a = k \times b}$ and this implies that $\mathbf{7 = 1 \times 7 + 0}$.

5.2 Exercise of application

1- Find separately subsets dividers (d) natural numbers **4, 6, 8,** and **10,** and then deduce the set (D) of their common divisors.

Solution:

Denoting by D_1 subset divisors of 4, it is noted that it is divisible by 1, 2 and 4 there will therefore be:

$4 \div 1 = 4, r = 0 \leftrightarrow 4 = (1 \times 4) + 0$ or $4 = 4 \times d_1$ with $d_1 = 1$

$4 \div 2 = 2, r = 0 \leftrightarrow 4 = (2 \times 2) + 0$ or $4 = 2 \times d_2$, with $d_2 = 2$

$4 \div 4 = 1, r = 0 \leftrightarrow 4 = (4 \times 1) + 0$ or $4 = 4 \times d_4$, with $d_4 = 4$.

There will therefore be the subset of integer divisors only 4, $D_1 = \{1, 2, 4\}$. For the integer 6 we note that:

$6 \div 1 = 6, r = 0 \leftrightarrow 6 = (1 \times 6) + 0$ or $6 = 6 \times d_1$ with $d_1 = 1$

$6 \div 2 = 3, r = 0 \leftrightarrow 6 = (2 \times 3) + 0$ or $6 = 3 \times d_2$ with $d_2 = 2$

$6 \div 3 = 2, r = 0 \leftrightarrow 6 = (3 \times 2) + 0$ or $6 = 3 \times d_3$ with $d_3 = 3$

$6 \div 6 = 1, r = 0 \leftrightarrow 6 = (6 \times 1) + 0$ or $6 = 6 \times d_6$ with $d_6 = 6$.

The subset of divisors of the natural number 6 is $D_2 = \{1. 2, 3, 6\}$. For the integer 8 we note that:

$8 \div 1 = 8, r = 0 \leftrightarrow 8 = (1 \times 8) + 0$ ou $8 = 8d_1$ avec $d_1 = 1$

$8 \div 2 = 4, r = 0 \leftrightarrow 8 = (2 \times 4) + 0$ ou $4 = 2d_2$ avec $d_2 = 2$

$8 \div 4 = 2, r = 0 \leftrightarrow 8 = (4 \times 2) + 0$ ou $4 = 4k$ avec $d_4 = 4$

$8 \div 8 = 1, r = 0 \leftrightarrow 8 = (8 \times 1) + 0$ ou $4 = 4k$ avec $d_8 = 8$

This gives the subset of integer divisors only 8 $D_3 = \{1. 2, 4, 8\}$. For the natural number 10 we note that:

$10 \div 1 = 10, r = 0 \leftrightarrow \mathbf{10 = (1 \times 2) + 0}$ or $\mathbf{10 = 10 \times d_1}$, with $\mathbf{d_1 = 1}$

$10 \div 2 = 5, \ r = 0 \leftrightarrow \mathbf{10 = (2 \times 5) + 0}$ or $\mathbf{10 = \ 5 \times d_2}$, with $\mathbf{d_2 = 2}$

$10 \div 5 = 2, \ r = 0 \leftrightarrow \mathbf{10 = (5 \times 2) + 0}$ or $\mathbf{10 = 2k}$, with $\mathbf{d_8 = 5}$

$10 \div 10 = 1, r = 0 \leftrightarrow \mathbf{10 = (10 \times 1) + 0}$ or $\mathbf{10 = 10k}$, with $\mathbf{d_{10} = 10}$ on have, $\mathbf{D_4 = \{1. 2, 5, 10\}}$.

2 - we see that the common divisors associated with subsets $\mathbf{d_1, d_2, d_3, d_4}$ are $\mathbf{d_1, d_2}$, we deduce then the set of common divisors D, is $\mathbf{D = \{1, 2\}}$.

6 dividers and multiples

6.1 Dividers

- Definition

We say that the integer **b** is divisor of the integer **a**, if and only if, the remainder **r** of the Euclidian division of **a** by **b**, is zero, **r = 0**, or **a** is a multiple of **b**.

6.1.1 The non-common divisor

Is called a non-common divisor of any two integers or more, the natural number **n**, which divides one, without dividing the other.

6.1.2 The common divisor

Called a common divisor of a subset **E** of natural numbers, the integer number n, which simultaneously divides the natural subset of integers **E.**

Example

Find non common divisors and the common divisors of the set n, of integers natural E = {4, 6, 8}.

- Solution

We writes :

4 = 1× 2 × 2

6 = 1× 2 × 3

8 = 1× 2 × 4 = 2 × 2 × 2.

On note that :

1 divide together **4, 6, 8**

2 divide **4, car 4 ÷ 2 = 2 reste 0**

4 divide 4.

It will then have the subset divisors of 4, d1 = {1, 2, 4}, We also note that:

1 divides **6**

2 divides **6**

3 divides **6**

6 divides **6.**

There will therefore be the subset d2 dividers 6, d2 = {1, 2, 3, 6}.

Whether d3, the set of dividers 8.

We also note that.

1 divides 8

2 divides 8

3 does not divide 8 because the rest is 2, r = 2

4 divides 8

8 divides 8

There will therefore be the subset d3 dividers 8, d_3 = {1, 2, 4, 8}

on the one hand it is noted that 3 divides 6, and does not divide 4 and 8. On the other hand, 4

divides 4, and does not divide 6 and 8.

We also note that 6 divides 6, and does not divide 4 and 8. On the other hand, 8 divides 8, and does not divide 4 and 6.

And it follows that, all non-common divisors E is D_0, one writes:

$D_0 = \{3, 4, 6, 8\}$.

as deduced, the set D, of common divisors $E = \{4, 6, 8\}$.

One writes: $D_c = \{1, 2\}$.

- **Application 2**

Be the subset of integers $N_1 = \{12, 14, 24, 28, 42, 72, 56, 60, 70\}$. Find all non-common divisors D_1 subset N_1 and deduce the common divisors of N_1.

The number 12

We begin by performing dividing 12 by 2, we get 6 remains 0, we writes:

$12 = 2 \times 6 \rightarrow 1$

Then performs the successive division of the Euclidean division of 6 by 2 gives 3 is the remainder is 0

$6 = 2 \times 3 \rightarrow 2$

Product 2×3 is replaced in operation 1, we have:

$12 = (2 \times 2) \times 3 \times 3 = 4$

It is concluded that the dividers 12 are: $d = \{1, 2, 3, 4, 6, 12\}$.

The number 14

Is carried out the successive division of the Euclidean division of 14 by 2, we obtained 7 is the remainder is 0.

$14 = 1 \times 2 \times 7 \times 14 = 1$.

It is concluded that the dividers 14 are: **d** = {1, 2, 7, 14}.

The number 24

It performs the successive division of the Euclidean division of 24 by 2, we get 12, remains 0, one writes:

$24 = 2 \times 12 \rightarrow$ **3**

12 and then divided by 2, we obtain:

$12 = 2 \times 6 \times 2 = (2 \times 3)$

We replaces produces 2×6 in step 3, we will have:

$24 = 2 \times 2 \times 6 \times 4 = 6$

$6 = (2 \times 3)$

We concluded that the dividers 24 are: **d** = {1, 2, 3, 4, 6, 8, 12, 24}.

The number 28

Then performs the successive division of the Euclidean division of 28 by 2, we obtain 14 remains 0, then:

$28 = 2 \times 14 \rightarrow$ **4**

14 and then divided by 2, we obtain:

$14 = 2 \times 7$

We replace the product 2×7 in step 4, we will have:

$28 = 1 \times 2 \times (2 \times 7) = 2 \times 14 = (2 \times 2) = 1 \times 7 \times 4 \times 1 \times 7 = 28$.

We conclude that the dividers of 28 are: $d = \{1, 2, 4, 7, 14, 28\}$.

The number 42

We carry out the successive division of the Euclidean division of 42 by 2, 21 is obtained, the remainder 0, one writes:

$42 = 2 \times 21 \rightarrow \mathbf{5}$

21 and then divided by 3, we get:

$21 = 3 \times 7$

Then 21 is replaced by its product 3×7 in step 5, we will have:

$42 = 1 \times 2 \times (3 \times 7) = (2 \times 3) \times 7 \times 7 = 6$

We conclude that the dividers 42 are: $d_2 = \{1, 2, 3, 6, 7, 21, 42\}$.

The number 72

Is carried out the successive division of the Euclidean division of 72 by 2 is obtained 36 and the remainder is 0.

$72 = 2 \times 36 \rightarrow \mathbf{6}$

36 and then divided by 2, we obtain:

$36 = 2 \times 18$

36 is replaced by its product 2×18 6 in the operation, we have:

$72 = 2 \times (2 \times 18) \rightarrow \mathbf{7}$

18 is divided by 2, we obtain:

$18 = 2 \times 9$

18 is replaced by its product 2×9 in step 7, we have:

$72 = 2 \times 2 \times (2 \times 9) = (2 \times 2 \times 2) = 8 \times 9 \times 9 \rightarrow \mathbf{8}$

9 then divided by 3 gives:

$9 = 3 \times 3$

Then 9 is replaced by its product 3×3 in step 8, we will have:

$72 = 1 \times [(2 \times 2) \times 2] \times 3 \times 3 \times 1 = (2 \times 3) \times (4 \times 3) = 6 \times 12 \times 4 = 18 = 24 = 36 \times 3 \times 2$

We conclude that the dividers 72 are: $\mathbf{d_3} = \{1, 2, 3, 4, 6, 8, 9, 12, 18, 24, 36, 72\}$.

The number 60

Is carried out the successive division of the Euclidean division of 60 by 2, 30 is obtained and the rest 0.

$60 = 2 \times 30 \rightarrow$ **11**

Then we divide 30 by 2 we get:

$30 = 2 \times 15$

30 is replaced by its product **2 × 15** in operation 11, we will have:

$60 = 2 \times (2 \times 15) \rightarrow$ **12**

15 and then divided by 3, since it is not divisible by 2, we obtain:

$15 = 3 \times 5$

Then 15 is replaced by its product **3 × 5** in Operation 12, we will have:

$60 = 1 \times (2 \times 2) \times (3 \times 5) = 4 \times 15 \times 10 = 6$

We conclude that the dividers 56 are: $d_4 = \{1, 2, 3, 4, 5, 6, 10, 15, 60\}$.

<u>The number 70</u>

And then performs the successive operation of the Euclidean division of 70 by 2, 35 is obtained and the rest 0.

$70 = 2 \times 35 \rightarrow \mathbf{13}$

Then we divide 35 by 5 we get:

$35 = 5 \times 7$

Then 35 is replaced by the product 5×7 in the operation 13, we will have:

$70 = 1 \times 2 \times (5 \times 7) = 10 \times 7 \times 2 = 35$

We conclude that the dividers 70 are: $d5 = \{1, 2, 5, 7, 10, 35, 70\}$.

All operations are, it is deduced that all non-common divisors of N1 is: $D0 = \{1, 2, 3, 4, 5, 6, 7, 10, 14, 15, 28, 42\}$, and all common factors $\mathbf{N_1}$ is: $\mathbf{D_c} = \{1, 2\}$.

6.2 The multiples

A natural number **a,** is a multiple of the integer **b**, if only the remainder of the Euclidean division of **a** by **b** is zero, **r = 0.**

Applications:

1- $4 \times 2 = 2$, we say that 4 is a multiple of 2, or 2 is a divisor of 4

$6 = 2 \times 3$, we say that 6 is a multiple of 3, or 3 is a factor of 6.

Is obtained throughout the several natural numbers of 2, by successively multiplying it by 2, 3, 4, 5, ... n, we obtain M = {2, 6, 8, 10, 12, 14, 16, 18, 20, 22, 24, ...n}.

Find n natural numbers, multiples of 3, and lower than 20, $n \leq 20$.

We design by m, all these multiple integers of 3 are: m = {3, 6, 9, 12, 15, 18}.

2 - Find all natural numbers n M, multiples of 2, checking, $10 \leq n \leq 30$.

Solution:

We note that **10** is a multiple of **2**, because $10 = 2 \times 5$, and if we search for natural numbers, multiples of 2, between 10 and 30, we find M : = {10, 12, 14, 16, 18, 20, 22, 24, 26, 28, 30} of all even numbers.

6.3 Divisibility by 2, 3, 4, 5, 6, 7, 8, 9

6.3.1 The divisibility by 2

• **Theorem 1**

A natural number a, is divisible by **2**, if and only if, there is an even number, or figure of the unit is an even number or zero

4 is an even number, it is divisible by 2.

8 is an even number, it is divisible by 2

10 is an even number, the number of the unit is zero, it is divisible by 2

16 its sales of the unit is even, and this is an even number, it is divisible by 2.

6.3.2 Divisibility by 3

• Theorem 2

A natural number **a**, is divisible by **3**, if and only if, it is a multiple of **3**, or the sum of its digits is a multiple of 3.

Example

6 is divisible by 3, because it is a multiple of 3.

9 is divisible by 3, because it is a multiple of 3.

15 is divisible by 3, because the sum of the digits is, $1 + 5 = 6$, which is a multiple of 3.

18 is divisible by 3, because the sum of its digits is $8 + 1 = 9$, and 9 is a multiple of 3.

21 is divisible by 3, because the sum of its digits is $2 + 1 = 3$, and 3 is divisible by 3.

6.3.3 The divisibility by 4

• Theorem 3

A natural number **a**, is divisible by **4**, if and only if, the number constituting its units and hundreds, is divisible by **4**, or is a multiple of **4**.

104 is divisible by 4, since 04 is divisible by 4

116 is divisible by 4, since 16 is divisible by 4, and a multiple of 4

124 is divisible by 4, since 24 is divisible by 4, and it is a multiple of 4.

6.3.4 The divisibility by 5

• **Theorem 4**

A natural number **a**, is divisible by 5, if and only if, its sales of the unit is a 0 or 5.

Example 2

10 is divisible by 5 because its unit is 0

35 is divisible by 5 because his unit 5

80 is divisible by 5 because its unit is 0

125 is divisible by 5 because his unit 5.

6.3.5 The divisibility by 6

• **Theorem 5**

A natural integer **a** is divisible by **6**, if and only if, it is a multiple of 6, or simultaneously divisible by **2** and by **3**

Examples

18 is divisible by 6, because 18 is evenly divisible by 2, and it is a multiple of 3, actually,

8 + 1 = 9 and 9 is divisible by 3.

24 is divisible by 6 because 24 is divisible by 2 and 3

36 is divisible by 6, because 36 is evenly divisible by 2, and this is a multiple of 3, actually,

3 + 6 = 9 and 9 is divisible by 3.

72 is divisible by 6 because 72 is divisible by 2, and is a multiple of 3, actually,

7 + 2 = 9 and 9 is divisible by 3.

6.3.6 The divisibility by 7

• **Theorem 5**

A natural number **a**, is divisible by **7**, if and only if, it is a multiple of 7

Example

14 is divisible by 7 because it is a multiple of 7

21 is divisible by 7 because it is a multiple of 7

98 is divisible by 7 because it is a multiple of 7

6.3.7 The divisibility by 8

• **Theorem 5**

A natural integer a is divisible by 8, if and only if it is a multiple of 8, or is simultaneously divisible by 2 and 4.

Example:

24 is divisible by 8 because 24 is divisible by 2, and is a multiple of 4, in fact,

$6 \times 4 = 24$.

48 is divisible by 8 because 48 is divisible by 2, and is a multiple of 4, in fact, $12 \times 4 = 48$.

6.3.8 The divisibility by 9

A natural number a, is divisible by 9 if and only if it is a multiple of 9, or the sum of its digits is a multiple of 9.

Example

27 is divisible by 9 because the sum of its digits is $2 + 7 = 9$, which is divisible by 9

45 is divisible by 9, because the sum of the digits is $4 + 5 = 9$, which is divisible by 9

54 is divisible by 9, because the sum of the digits is $5 + 4 = 9$, which is divisible by 9

72 is divisible by 9 because the sum of its digits is $7 + 2 = 9$, which is divisible by 9

6.3.9 The divisibility by 10

> A natural number **a**, is divisible by **10**, if and only if, it is a multiple of 10, or simultaneously it's divisible by 2 and 5, or the number of the unit is a zero 0.

20 is divisible by **5** because its unit is **0**, and simultaneously it is divisible by **2** and by **5**

30 is divisible by **5** because its unit is **0**, and simultaneously it is divisible by **2** and by **5**

70 is divisible by **5** because its unit is **0**, and simultaneously it is divisible by **2** and by **5**.

Property :

-        If a natural integer **a** is divisible by separately integers **b** and **c**, it is divisible by their product, **b × c.**

Examples:

8 is divisible by **2**

8 is divisible by **4**.

So: 8 is divisible by **2 × 4.**

42 is divisible by **6**

42 is divisible by **7**

Therefore: **42** is divisible by **6 × 7.**

Reciprocally :

> If a natural integer **a** is divisible by the product **b × c** natural numbers **b** and **c**, it is divisible by separately, **b** and **c**.

42 is divisible by **42**, or **42 × 7 = 6**, so:

42 is divisible by **6** and **7**

6.4 The greatest common divisor (GCD)

Example

We denote by E_0 the set of integers n, $E_0 = \{2, 4, 6, 8, 10\}$, find the **GCD** of E_0.

To find the **GCD** of E_0, we first divide each number by 2, by following the following method:

$2 = 2 \times 1$

$4 = 2 \times 2$

$6 = 2 \times 3$

$8 = 2 \times 4$

$10 = 2 \times 5$

- Find the greatest common divisor (**GCD**) is to seek the greatest integer that divides simultaneously, 2, 4, 6, 8, 10

We Note that 2 divides simultaneously 2, 4, 6, 8, 10 is their **GCD**.

So the **GCD** = 2

Find the greatest common divisor (**GCD**) of 15, 21, 24, 36.

We write :

$15 = 3^1 \times 5$

$21 = 3^1 \times 7$

$24 = 2 \times 2 \times 2 \times 3 = 2^3 \times 3^1$

$36 = 2 \times 3 \times 2 \times 3 = 2^2 \times 3^2.$

We obtain **le PGCD** $= 2^2 \times 3^1 = 4 \times 3 = 12.$

6.4.1 Theorem

we obtain the **GCD** a set of integers, by obtaining the product of the common factors of all numbers including, smaller exponents.

6. 5 the least common multiple (LCM)

Example

E_1 means all natural numbers n, $E_1 = \{2, 4, 6, 8, 10\}$, find the **LCM** of E_1.

To find the **LCM** of **E_1,** we first divide each number by Euclidean division by 2, 3, 5, 7, and the following method

$2 = 2 \times 1$

$4 = 2 \times 2 = 2^2$

$6 = 2 \times 3$

$8 = 2 \times 4 = 2 \times (2 \times 2) = 2^3$

$10 = 2 \times 5$

- Find the least common multiple (**LCM**) is seeking the smallest integer that divides simultaneously, 2, 4, 6, 8, 10.

The **LCM** of these numbers is to find their common factors and not shared with the greatest exponent, and put into product, from the previous successive division, we have:

The **LCM** $= 1 \times 2^3 \times 3 \times 5 = 8 \times 15 = $ **120**.

Find the least common multiple (**LCM**) of 15, 21, 24, 36.

We write :

$15 = 3 \times 5$

$21 = 3 \times 7$

$24 = 2 \times 2 \times 2 \times 3 = 2^3 \times 3$

$36 = 4 \times 9 = 2 \times 2 \times 3 \times 3 = 2^2 \times 3^2$.

The **LCM**, should include common and not common factors with the greatest exponent.

So we have: **LCM** $= 2^3 \times 7 \times 5 \times 3^2 = (2 \times 2 \times 2) \times (3 \times 3) \times (5 \times 7) = 8 \times 9 \times 35 = 2520$

LCM = 2520.

Verification:

2520 is the least common multiple of 15, 21, 24, 36, this means that **2520** is the small number divisible shared simultaneously by 15, 21, 24, 36, parallel to the largest common multiple that will: **GCM** $= 15 \times 21 \times 24 \times 36 = 272{,}160$; **GCM = 272160**.

Note

In arithmetic and algebra and analysis, we work with the LCM, which gives us a simplified number, while **GCM** gives us a big no simplified, even all both lead to the same result.

6.5.1 Theorem

> We obtain the **LCM** of a set of natural numbers as the product of the common factors and not common to all these numbers, and including the greatest exponent.

7 Exercise resolved

7.1 Exercise

Find the **GCD** and **LCM** of the set of natural numbers $N_0 = \{10, 15, 24, 27, 32, 57\}$.

Solution :

Is denoted by Dc the **GCD** of all N_0, and is the Euclidean division by 2, 3, 5, 7 natural numbers N_0, we write:

$10 = 1 \times 2 \times 5$

$15 = 1 \times 3 \times 5$

$24 = 1 \times 2 \times 12 = 2 \times 2 \times 6 = 2 \times 2 \times 2 \times 3 = 2^3 \times 3$

$27 = 1 \times 3 \times 9 = 3 \times 3 \times 3 = 3 \times 3 \times 3 = 3^3$

$32 = 1 \times 2 \times 16 = 2 \times 2 \times 8 = 2 \times 2 \times (2 \times 4) = 2 \times 2 \times (2 \times 2 \times 2) = 2^5$

$57 = 1 \times 3 \times 19.$

The **GCD**, which is the product of the common factors of all N_0, including the smallest exponent. Note that 1 is the only divisor of these numbers, and there is no other common factor between them, which leads us to say:

The **GCD = 1**.

For cons, the **LCM** embedded common and not common factors with the greatest exponent, we shall have:

LCM $= 25 \times 33 \times 5 \times 19 = 82080.$

8 Index

8.1 Mathematical symbols

symbols	Signification	examples
=	equal	$2 = 2$
≠	different	$3 \neq 2$
>	better than	$2 > 1$
<	less than	$2 < 3$
≤	less than or equal	$n \leq 10$
≥	greater than or equal	$n \geq 12$
∈	belongs to	$2 \in \{1, 5, 2, 4, 6\}$
∪	union of sets	$\{1, 5, 2, 4\} \cup \{2, 7, 2, 4\}$
N	sets of natural numbers including zero	$N = \{0, 1, 2, 3, 4, 5, \dots n\}$
N*	sets of natural numbers excluding zero	$N^* = \{1, 2, 3, 4, 5, \dots n\}$

Even partial reproduction of this manuscript, it may be its version, written, video, or in translation, is strictly prohibited without prior authorization from the author.

The civil code punishes those who brought harm to the intellectual property of others.

www.ingramcontent.com/pod-product-compliance
Lightning Source LLC
Chambersburg PA
CBHW050832180526
45159CB00004B/1870